C000234117

Men-at-Arms • 217

The War in Laos
1960–75

Kenneth Conboy • Illustrated by Simon McCouaig

Series editor Martin Windrow

First published in Great Britain in 1989 by Osprey Publishing,
Midland House, West Way, Botley, Oxford OX2 0PH, UK
44-02 23rd St, Suite 219, Long Island City, NY 11101, USA
Email: info@ospreypublishing.com

© 1989 Osprey Publishing Ltd.

All rights reserved. Apart from any fair dealing for the purpose of private study, research,
criticism or review, as permitted under the Copyright, Designs and Patents Act, 1988,
no part of this publication may be reproduced, stored in a retrieval system, or transmitted
in any form or by any means, electronic, electrical, chemical, mechanical, optical,
photocopying, recording or otherwise, without the prior written permission of the
copyright owner. Enquiries should be addressed to the Publishers.

Transferred to digital print on demand 2010

First published 1989
3rd impression 2005

Printed and bound in Great Britain

A CIP catalogue record for this book is available from the British Library

ISBN: 978 0 85045 938 8

Series Editor: Martin Windrow
Filmset in Great Britain

Artist's note

Readers may care to note that the original paintings from which the colour plates in this book were prepared are available for
private sale. All reproduction copyright whatsoever is retained by the Publishers. All enquiries should be addressed to:

Simon McCouaig
4 Yeoman's Close
Stoke Bishop
Bristol
BS9 1DH
UK

The Publishers regret that they can enter into no correspondence upon this matter.

FOR A CATALOGUE OF ALL BOOKS PUBLISHED BY
OSPREY MILITARY AND AVIATION PLEASE CONTACT:

Osprey Direct, c/o Random House Distribution Center,
400 Hahn Road, Westminster, MD 21157
Email: uscustomerservice@ospreypublishing.com

Osprey Direct, The Book Service Ltd, Distribution Centre,
Colchester Road, Frating Green, Colchester, Essex, CO7 7DW
Email: customerservice@ospreypublishing.com

www.ospreypublishing.com

The Course of the War

Modern Laos began as a peripheral protectorate of French Indochina—a buffer state designed to shield greater French interests in northern Vietnam from both the Kingdom of Thailand and the British in Burma. With a population of only three million people, Laos was fragmented into three major ethnic groups. The lowland Lao, who made up less than half the total population, worked closely with the French and dominated the country's budding administrative services and military. Lao Theung tribesmen, concentrated in the southern Laotian panhandle, comprised 25 per cent of the population. The Hmong hill tribes, who were the last to migrate to Laos from present-day southern China, comprised about five per cent.

From 1893 until World War Two, Laos remained a relatively peaceful French protectorate. Its fractured ethnic composition gave Laos little sense of a national identity, enabling the Imperial Japanese military to seize control in mid-1945 without encountering any significant resistance. As the war drew to a close, the Japanese encouraged the growth of indigenous nationalist movements to forestall the return of French power to mainland South-East Asia. Despite these efforts the French re-entered Indochina, swept away the small Free Lao nationalist movement, and methodically retook the protectorate by 1946.

Soundly defeated, the Free Laos moved to Thailand. In 1949 a splinter faction led by Prince Souphanouvong made its way into northern Vietnam and contacted the anti-French Viet Minh forces. This group became the core of the

Capt. Kong Le, commander of the 2ᵉ Bataillon de Parachutistes, August 1960. He wears French airborne camouflage clothing with US pistol belt, and khaki peaked cap with gold ANL cap badge. Above the metal basic parachutist's wings are his rank insignia, three gold stars on a rectangle of red cloth. (Courtesy Frank Tatu)

Communist Laotian forces, known commonly as the Pathet Lao. Trained and equipped by the Viet Minh, Souphanouvong's men concentrated on infiltrating the north-east Laotian province of Sam Neua. In 1953 the Viet Minh launched a multi-divisional invasion of northern Laos, crushed the

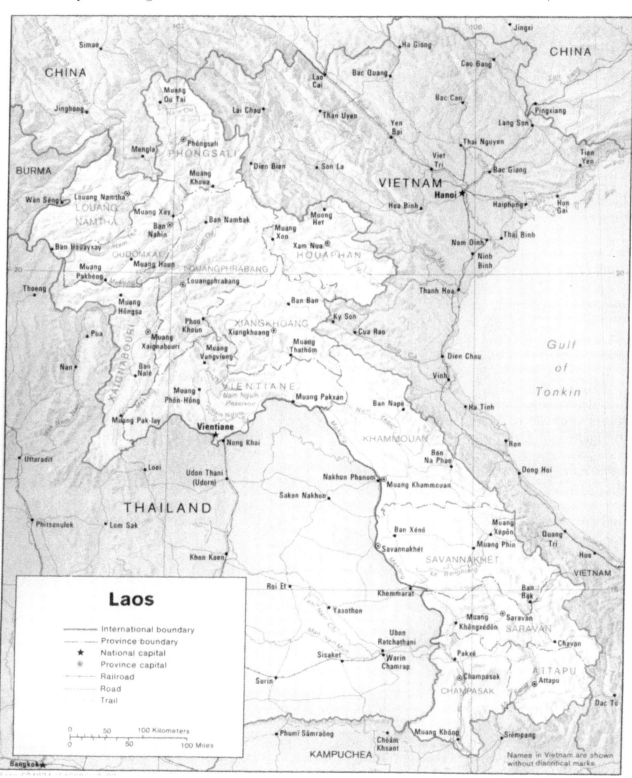

French presence in Sam Neua, and allowed the Pathet Lao to move in and establish their own political and military headquarters in the province.

1957: The Abortive Coalition

Following the 1954 Geneva Agreements calling for the withdrawal of French military forces from Indochina, the Armée Nationale de Laos—the armed forces of the independent Kingdom of Laos—continued to fight the Pathet Lao. After three years of negotiations, an agreement for the formation of a coalition government was signed in November 1957. The political wing of the Pathet Lao was to be recognised as a legitimate party, and the Pathet Lao fighting units were scheduled to be integrated into the ANL over the next two years.

The coalition government showed signs of strain almost as soon as it was formed. Although some Pathet Lao were symbolically incorporated into the ANL in February 1958, further integration was stalled. Finally, in May 1959, two Pathet Lao battalions were brought in from the field. At their integration ceremony, however, one battalion fled and crossed the border into North Vietnam. That same month, Pathet Lao units supported by the North Vietnamese Army staged a foray into Sam Neua province. The government, shaken by its initial reverses, launched a counterattack, and gradually retook the lost positions by autumn.

1960: Kong Le's Coup

As 1959 ended the ANL undertook a wider offensive against the Pathet Lao, shuttling back and forth across the country and turning the tide against the Communist insurgents. Most active in the government offensive was the crack 2ᵉ Bataillon de Parachutistes led by Capt. Kong Le. On 9 August 1960, after several weary months of combat, Kong Le took matters into his own hands. While most government dignitaries were in Luang Prabang attending the king's funeral, the 2ᵉ BP swept across Vientiane, the capital of Laos. Kong Le then announced the formation of a neutralist government which would be open to both the Royalists and the Communist Pathet Lao.

The Pathet Lao took immediate advantage of Kong Le's offer, and began sending their forces into Vientiane. The Soviets assisted by airlifting in supplies and an NVA artillery battery. The Royalists, with the growing approval of the US, grouped in the southern town of Savannakhet and prepared for a counterattack on Vientiane. Led by Gen. Phoumi Nosavan, the Savannakhet forces launched a combined ground and airborne assault in early December. Kong Le's Neutralist paratroopers and the Pathet Lao withdrew north from the capital in an organised fashion, then veered east, and conquered the strategic Plaine des Jarres in central Xieng Khouang Province by New Year's Eve.

1962: The Second Coalition

Once secure in Vientiane, the ANL repeatedly delayed an all-out attack to recapture the Plaine des Jarres. By May 1961 a ceasefire was signed and hostilities dropped off. The Laotian armed forces, renamed the Forces Armées du Royaume, concentrated on small-scale sweeps until heavy fighting broke out again in February 1962 when the Communists began exerting heavy pressure on the north-western garrison at Nam Tha. Neither parachute reinforcements from the FAR's Groupement Mobile 15, nor T-6 bomber support were able to silence the enemy positions in the hills around Nam Tha. On 5 May the garrison began to collapse; over 2,000 FAR troops headed for the border, many not stopping until they had crossed the Mekong River into Thailand.

With the fall of Nam Tha, international attention was focused on Laos. President Kennedy ordered 5,000 soldiers to Thailand to emphasise US commitment to the Royalist government. By June fighting subsided; and a second coalition government was formed between the Royalists, Neutralists, and Pathet Lao Communists. Complying with the new rules of Laotian neutrality, the US withdrew its entire Military Advisory and Assistance Group by 6 October 1962. The NVA, who had increased their assistance to the Pathet Lao while bolstering their forces along the Ho Chi Minh Trail running down the Laotian Panhandle, responded by removing only a handful of 'technicians'.

1964

Under the new tripartite government, peace in Laos lasted for six months. By 1963 relations between the Pathet Lao and Kong Le Neutralists

5

began to fray. Clashes flared across the Plaine des Jarres, and in May 1964 a concerted Pathet Lao offensive swept across the strategic plain. Kong Le gathered his remaining loyal paratroopers and turned back to the Royal Government for support. Mindful of Kong Le's deceit in 1960, the FAR nevertheless fell into a loose alliance with the Neutralists. While maintaining the façade of a separate army, the Neutralists were effectively reduced to a subordinate branch of the FAR.

In July, reacting to Pathet Lao pressure north of the capital, the government launched Operation 'Triangle', a three-pronged FAR and Neutralist offensive against the crossroads town of Sala Phou Khoun. After two weeks of fluid fighting, 'Triangle' succeeded in routing Pathet Lao forces along the highway between Vientiane and Luang Prabang. Contributing heavily to the success of the operation was the extensive use of tactical air support: the fledgling Royal Laotian Air Force, augmented by volunteer Thai pilots, flew repeated T-28 sorties in support of the three task forces. Operation 'Triangle' also marked the first time a regimental-sized FAR unit was airlifted during the rainy season. This set a precedent for major government offensives during the wet season in the years to come.

1965

As the year opened fighting shifted from the countryside to the capital, and a series of rightist coups and counter-coups shook Laos. While political skirmishes flared in Vientiane the war in the countryside was entering a new stage. Fighting was concentrated in the northern portion of the country, where the Pathet Lao seized the initiative in the dry season lasting from October to May. They travelled in platoon-sized elements, holding few static positions in true guerrilla fashion. The government forces pulled back to a string of fortified bastions surrounding the Plaine des Jarres, leaving tactical airpower to hunt the elusive enemy forces. Once the monsoons arrived in June the

Hmong ADC guerrillas accompany the crew of a civilian contract airliner searching for a crash site in northern Laos, late 1961. The Hmong wear US OG 107 fatigues. A Thai PARU advisor standing in the centre is wearing a black beret with metal Thai police wings used as a beret badge. (Courtesy Brig.Gen. Harry Aderholt)

Pathet Lao were forced to withdraw toward Sam Neua as their resupply network bogged down. Taking advantage of superior aerial resupply assets, the government moved into the vacuum left by the Pathet Lao and leapfrogged deep into enemy territory. Fighting was rarely intense, and few casualties were suffered by either side. One noticeable shift that occurred on the government side was that the war in the north was increasingly being fought by Hmong guerrillas under the command of Gen. Vang Pao. A rugged and fiercely independent minority, the Hmong proved themselves adept at irregular warfare, in contrast to the characteristically unspectacular performance of the FAR.

In the south the war was being fought at the same sporadic tempo. Small government intelligence, roadwatching, and action teams were raised in 1965 to begin operations against the NVA on the Ho Chi Minh Trail. In addition, two companies of Nung soldiers fighting in support of the government were based along the eastern Bolovens Plateau and used for larger scale diversionary attacks against the Vietnamese forces on the Trail.

1966–8: The NVA Initiative

In 1966 the annual pattern of fighting around the Plaine des Jarres entered its second year. By June, Hmong guerrillas had overwhelmed the remaining pockets of enemy forces and extended their operations virtually to the North Vietnamese border. The yearly shift across northeast Laos was repeated in 1967, with the government moving right up to the Pathet Lao headquarters in Sam Neau.

By this time, the NVA were firmly enmeshed in the war in South Vietnam, and were growing impatient with the lacklustre performance of their Pathet Lao allies. As a result, when the Communists initiated their dry season offensive in northern Laos during January 1968, it was the NVA who assumed control of the fight. Vietnamese forces smashed the government enclaves in the north east, culminating on 14 March with a daring sapper attack on a US Air Force radar facility perched high atop a seemingly impregnable mountain garrison. The radar base, known as Lima Site 85, was overrun with the loss of 13 American lives.

US Air Force Combat Weatherman from Detachment 75, Air Weather Service, poses with a Hmong guerrilla commander, 1965. The airman wears civilian clothes and carries an M-16 rifle. The guerrilla, an SGU battalion commander, carries M-59 grenades, and an M-79 grenade launcher.

Vang Pao withdrew his weary forces to his headquarters at Long Tieng, waiting until November to launch his counteroffensive, Operation 'Pigfat'. Used in their first major airmobile assault, the guerrillas stormed the base of the mountain containing Lima Site 85. Heavy fighting continued for a month, with airpower raking the sides of the ridge. The NVA took excessive casualties, but held; the Hmong were forced to retreat to Long Tieng.

1969: The NVA Over-Extend

Charging on the heels of Vang Pao's demoralised forces, the NVA renewed their offensive in February 1969. Surrounding the guerrilla stronghold at Nakhang, the NVA hit it with elements of three regiments. In previous years the Communists had allowed the government forces to 'escape' from isolated garrisons if they ceased resistance. This time, the Nakhang garrison was decimated piecemeal. Over-confident, the NVA continued with their offensive, extending their lines across the Plaine des Jarres and capturing the Neutralist post at Muong Soui on 28 June.

With their stretched forces encountering little resistance, the NVA made the fateful decision to hold on to their captured territory into the rainy season. It was a major mistake. Under the codename Operation 'Stranglehold', allied tactical airpower was immediately used to cut off the major NVA supply routes east of the Plaine. By July, intelligence reports indicated that the Vietnamese were running short of supplies and, in some cases, starving.

While the NVA was being heavily bombed, Vang Pao began to plan for an ambitious operation to recapture the entire Plaine des Jarres for the first

time since 1960. Codenamed Operation 'About Face', the offensive started in early August with pincer assaults to cut off the eastern access to the plain near the crossroads village at Ban Ban. Two guerrilla regiments then pushed on to the plain from the south, catching the NVA by surprise and capturing tons of food, medicine, and military hardware. Some 25 Soviet PT-76 tanks were taken, along with 113 other assorted vehicles.

While 'About Face' was in full swing, an equally ambitious operation named 'Junction City' was being launched in the south from Savannakhet. Four battalions of Lao guerrillas captured the regional Pathet Lao headquarters at Muong Phine in October. Guerrilla patrols then pushed on right to the outskirts of the major NVA transshipment point at Tchepone. Although Communist forces retook the town later in the month, 'Junction City' proved that well-defended targets could be defeated by fast-moving irregular forces. From that point on the Royal Laotian government had

Hmong guerrilla poses next to a red plastic arrow used to direct air strikes around the Plaine des Jarres, 1965. The guerrilla wears OG 107 fatigues, with M-59 grenades on a US pistol belt.

become, in effect, guerrillas in their own country, while the Communist enemy was a conventional force.

1970

Stung by its major defeat on the Plaine des Jarres, the NVA prepared for a major offensive in early 1970. The 312 and 316 Divisions massed north-east of the Plaine des Jarres, and an advance force of sappers and tanks stormed across the plain and overran the guerrilla lines. NVA infantry then pushed south-west of the plain, burning down the town of Sam Thong and threatening the adjacent guerrilla headquarters at Long Tieng. Reinforcements had to be flown in from around the country and from Thailand to reinforce Long Tieng. In late June a diversionary attack, Operation 'Leapfrog', was launched on the southern edge of the Plaine des Jarres. The enemy pulled back, and Long Tieng was saved for that year.

In the south, the expanding guerrilla forces at Savannakhet launched two more drives toward Muong Phine in an attempt to repeat the success of 'Junction City'. The first of these offensives, Operation 'Maengda', bogged down in July without attaining the desired results. The second push, known as Operation 'Tchepone', involved two guerrilla task forces which approached within five kilometres of the NVA-held town of Tchepone. Although falling short of its target, the northern task force discovered and destroyed a major Vietnamese logistical command post.

On the Bolovens Plateau in central Saravane Province, the NVA was in the midst of a campaign to create a wider buffer zone for the Ho Chi Minh Trail. This was because the change of governments in Cambodia in March 1970 had deprived Hanoi of the use of its major supply route through the Cambodian port of Sihanoukville, leaving only the Ho Chi Minh Trail to carry supplies to the south. Not wanting to risk any threats to their remaining supply artery, the NVA began to systematically eliminate government positions along the eastern Bolovens. An attempt to push back the NVA late in the year using two battalions of Cambodian volunteers ended in failure, forcing the Laotian government to postpone another drive to reclaim the plateau until adequate Lao forces could be assembled.

As 1970 drew to a close, two of Vang Pao's regiments, well rested after their military reverses earlier in the year, were helicoptered south of Ban Ban in November for a month of harassing operations. This helped sap the momentum of the Communist forces, which refrained from launching a major push toward Long Tieng base during the 1970–1971 dry season.

1971

Once the rains began in June 1971 Vang Pao recalled his far-flung forces to participate in the second offensive to recapture the Plaine des Jarres. By mid-July the operation was finished and the plain was back in government hands. This time, however, the NVA had withdrawn in an orderly fashion, leaving only a few scattered food caches to be taken by the advancing government forces. Intent on holding his positions, Vang Pao reinforced the Plaine des Jarres with artillery support bases manned by Thai volunteers. Four additional guerrilla regiments held blocking positions along the northern and eastern extremes of the plain. They then waited for the rains to end and for the NVA to make its move.

In southern Laos the push to recapture the Bolovens Plateau, Operation 'Sayasila', began on 28 July. The main target was the central town of Paksong, held by the 9th Regiment of the 968 NVA Division. Ground charges east of Pakse along Route 23 resulted in hundreds of casualties for the government forces. Seeking to regain momentum, a guerrilla regiment loaned from Savannakhet was air-assaulted nine kilometres south-east of Paksong. The last Vietnamese defenders were rooted from the devastated town by 15 September. As on the Plaine des Jarres, Thai artillery bases were established around Paksong to defend the Laotian positions inside the town.

Reacting to the Laotian advances on the PDJ and the Bolovens, the NVA planned for an early counteroffensive. Making extensive use of the massive Soviet 130mm gun, the NVA pounded the Thai bases on the Plaine des Jarres. With their 155mm howitzers outranged, the Thais took heavy casualties. When an armoured NVA thrust pushed on to the eastern plain, the government guerrilla forces fled. Worse, air support was impaired by foul weather and marauding Vietnamese MiG fighters.

On 20 December, after three days of ruthless fighting, the Plaine des Jarres fell once again.

Driving south-east, the NVA continued their offensive with an all-out assault on Long Tieng headquarters. Vietnamese units pushed their way on to an overlooking ridge and subjected the base to continuous heavy weapons fire; sapper teams even penerated as far as Gen. Vang Pao's house. With Long Tieng on the verge of collapse, reinforcements from Thailand and Savannakhet battled their way up the adjacent ridgelines and turned the tide. By March, with the support of B-52 strikes, the NVA was beaten back and Long Tieng was put out of direct danger for another year.

1972–73

Mirroring their offensive in the north, the NVA attacked Paksong in December 1971. The government forces were routed, sending Laotian and Thai defenders streaming toward the river town of Pakse. After the arrival of reinforcements from Savannakhet the government lines were stabilised half way between Pakse and Paksong. It was not until October 1972 that the government launched Operation 'Black Lion', the counteroffensive to retake the Bolovens. Airlifting in one Savannakhet regiment, and three Thai volunteer battalions north of Paksong, the task force pushed its way back into Paksong by 6 December. A Thai ground advance east along Route 23 then linked up with the Paksong forces on 23 December.

At the same time, two Pakse guerrilla regiments were air-assaulted in October 1972 near the town

Project 404 attaché talks with Hmong soldiers at Muong Hiem, 1966. The soldier in the centre wears lieutenant-colonel's rank insignia over his right shirt procket. The attaché wears commercial spotted camouflage. (Courtesy Col. Bill Keeler)

Brig.Gen. Sourith Don Sasorith, commander of the RLAF, poses in blue dress uniform, 1967. He wears a blue peaked cap with silver RLAF badge; collar insignia are for the FAR General Headquarters; the RLAF master pilot's wings are silver. (Courtesy Col. Paul Pettigrew)

of Saravane on the northern rim of the Bolovens. A third Pakse regiment was inserted into the Thateng region to the east, where it operated for three months before being beaten back by elements of the NVA 968 Division.

Back on the Plaine des Jarres, Vang Pao's exhausted Hmong guerrillas were barely able to contain the repeated NVA advances. Looking to relieve pressure on Long Tieng and secure a foothold on the plain before an anticipated ceasefire agreement, Vang Pao launched an understrength offensive. An entire Commando Raider Company was parachuted north of the Plaine des Jarres on the night of 14 August, and two heliborne guerrilla regiments linked up with them on the following morning. They drove south on to the plain, while other units launched harassing attacks to the east and south-east. Following poor weather and an accidental B-52 strike on friendly forces, the inconclusive mini-offensive sputtered to a close in September.

During November, with reinforcements from Thailand, Savannakhet, and north-western Laos, Vang Pao pushed north-east of Long Tieng. Driving against the southern rim of the Plaine des Jarres, the government forces were met by a swift NVA wave of infantry, PT-76 tanks, and Chinese K-63 armoured personnel carriers. The government forces withdrew in disarray, and the Plaine des Jarres, which had become one of the most embattled pieces of real estate in South-East Asia—remained in enemy hands.

On the Bolovens, the NVA waited until the eve of the Laotian ceasefire before launching their final landgrab. On 21 February NVA forces fought their way into Paksong, pushing out the defenders holding the town. B-52 strikes were flown against the Vietnamese attackers, but the government was unable to push back the NVA.

Once the 1973 ceasefire went into effect a coalition government, which had already failed in 1957 and 1962, was resurrected for the third time. Unlike previous occasions, however, the Communists now held the Plaine des Jarres, the Bolovens, and most of the countryside. Furthermore, the guerrilla units and Thai volunteers—the only effective forces available to the government—were being dissolved or integrated into the FAR.

Completely demoralised by their apparent abandonment by the Americans, the numerically superior FAR was slowly pushed back toward the Mekong River by an encroaching wave of Pathet Lao.

In May 1975, following the fall of South Vietnam and Cambodia, the FAR collapsed. The Pathet Lao moved into Vientiane and declared the establishment of a People's Democratic Republic in December. The third domino in Indochina had fallen.

Laotian Government Forces

The Royal Laotian Army

The Armée Nationale de Laos was created in 1952 as an outgrowth of the French colonial forces. In 1954 the French officers of the ANL were withdrawn; and the ANL was doubled in size when additional Laotian units in the French colonial forces were handed over to the Royal Laotian Government. Plagued by inexperienced leaders and a poor logistical network, the ANL never developed into a unified armed force. The country was divided into Military Regions which operated like autonomous fiefdoms; Military Region commanders effectively wielded more power than the ANL General Headquarters in Vientiane.

In 1960 the Kong Le coup forced the ANL to co-ordinate operations above the battalion level. By November, Gen. Phoumi Nosavan had organised the first ANL Groupement Mobile, the French equivalent of a regimental combat group. Originally a temporary tactical grouping, the GMs were converted in April 1961 into permanent organisations. Significantly, the GM commanders grew in power, often approaching the previously unchallenged influence of the Military Region commanders.

From 1962 to 1968 the Laotian army, renamed the Forces Armées du Royaume, was the primary force tasked with the defence of the Laos. An exception was in the north-eastern region, where operations were conducted by the Hmong irregular forces under Vang Pao. The primary FAR unit was the GM of three infantry battalions;

Thai volunteer pilot poses in front of a T-28 bomber after completing his hundredth mission over Laos, 1966. 'B Team' pilots were hired for six months and a minimum of one hundred sorties, most of them flown in support of Hmong guerrillas on the Plaine des Jarres. He wears a US flightsuit and survival vest without insignia. (Courtesy Col. Bill Keeler)

despite being designated for nationwide use, in practice each GM remained within a specific Military Region. Other FAR units included volunteer battalions, also assigned to a specific military region; regional battalions, similar to the volunteer battalions, and disbanded in 1965; and parachute battalions, organised into GMs and used as a mobile reserve.

In late 1967 the FAR sealed its fate by electing to defend its garrison at Nambac in northern Laos, an isolated post surrounded by Communist-held high ground. Five FAR regiments, including most of the country's airborne battalions and artillery, were committed to Nambac's defence. On 12 January 1968 the garrison fell. All of the units present were decimated, crippling the FAR and striking a heavy blow to the morale of the Laotian government. Following this débâcle the FAR was relegated to the static defence of population centres along the

Mekong River. In addition, all GMs were disbanded and replaced by independent battalions.

In 1971, with the Vietnamisation process in full swing in South Vietnam, a similar effort was attempted toward making the FAR a more effective, self-sufficient force. Following a US system of organisation, two FAR divisions were created. The First Division, based in Luang Prabang, was theoretically tasked with operations in northern Laos. The Second Strike Division, based in Seno, was oriented toward the south. By late 1974 a thinning of FAR ranks forced the two divisions to be replaced by a series of smaller, understrength brigades. These were maintained until May 1975, when the Pathet Lao entered Vientiane and dissolved the FAR.

Uniforms and Equipment

The basic uniform of the FAR was the khaki shirt and trousers, furnished initially by France and later by the United States. By the mid-1960s, US OG 107 utilities and local variants were adopted throughout the Laotian forces. A white dress uniform was used by officers.

In the field, the FAR used a wide variety of uniforms depending on availability from foreign aid sources. Camouflage was popular, especially among the paratroopers, who used the French 1947/51 and '47/52 camouflage uniforms. At least some of Kong Le's 2ᵉ BP also had British Denison smocks, originally supplied to the French during the Indochina War. One well-known FAR commander in 1965 went as far as painting his own camouflage directly onto olive drab fatigues. Black

---

Order of Battle, Forces Armées Royales, July 1962

Military Region One (North-western Laos)
Mobile Forces:
Groupement Mobile 11 (Luang Prabang)
GM 16 (Kiou Ka Cham)
Territorial Forces:
5 Bataillons Voluntaires [1]
4 Bataillons Regionales [2]
35 Auto Defense de Choc units

Military Region Two (North-eastern Laos)
Mobile Forces:
GM B (Pha Khao) [3]
GM 13 (Ban Man)
Territorial Forces:
2 Bataillons Voluntaires [1]
170 Auto Defense de Choc units [4]

Military Region Three (Upper Panhandle)
Mobile Forces:
GM 12 (Thakhek)
GM 14 (Mahaxay)
GM 15 (Seno) [5]
GM 17 (Muong Phalane)
Territorial Forces:
4 Bataillons Voluntaires [1]
1 Bataillon Regionale [2]
34 Auto Defense de Choc units

Military Region Four (Lower Panhandle)
Mobile Forces:
GM 18 (Pakse)
Territorial Forces:
4 Bataillons Voluntaires [1]
4 Bataillons Regionales [2]
33 Auto Defense de Choc units

Military Region Five (Capital zone)
Mobile Forces:
3 Bataillons Speciales [6]
Territorial Forces:
4 Bataillons Voluntaires
9 Auto Defense de Choc units

FAR Groupements Mobiles were composed of three battalions.

[1] Local battalions under permanent control of Military Region commander
[2] Merged with Bataillons Voluntaires in 1965
[3] Commanded by Lt.Col. Vang Pao
[4] Later combined with GM B to form core of guerrilla force in northern Laos
[5] Airborne regiment composed of 1, 11 and 55 Bataillons de Parachutistes
[6] DNC Airborne Regiment composed of 11, 33 and 99 Bataillons Speciales

Graduating class of RLAF pilots pose in front of a training aircraft at Udorn Air Base, 1968. Blue working uniforms are worn with blue RLAF peaked caps. (Courtesy Lt.Col. William Ritchie)

leather boots supplied by the US were common, although local copies of the French-inspired Bata boot were seen on occasion.

Copying the French paras, the red beret was adopted by the entire FAR. A wreathed trident in gold metal or yellow cloth was worn on the right side. A khaki field cap (or khaki peaked cap, for officers) was also issued. Later, a full range of jungle hats and baseball caps found their way into the FAR from South Vietnam.

Yellow and subdued nametags were occasionally worn above the right pocket. Plastic nameplates were worn on dress uniforms. Elite units like the Special Commando Company of the Second Strike Division had their unit name printed over their left pocket.

Rank insignia were direct copies of the French system until 1959, after which a new Laotian system was introduced. Officers wore red shoulderboards with gold borders and a gold wreathed trident at the inner end. Junior officers added an appropriate number of gold stars to the board; field grade officers had a single lotus leaf rosette plus an appropriate number of gold stars. General officers had a gold leaf design around the lower half of the board plus two or more silver stars. Enlisted grades wore cloth chevrons on the upper sleeve.

In the field, officers' shoulderboards were initially replaced by rectangular red cloth tabs bearing metal rank insignia and worn over the right shirt pocket. By the late 1960s, an American-style system was adopted in which metal, yellow cloth, or black subdued rank insignia were worn on the right collar. Branch insignia came in gold metal, yellow cloth, and black subdued cloth. They were worn on both collars if shoulderboards were worn, but only on the left collar if used together with rank collar insignia.

Like the South Vietnamese, the FAR were given to creating unit insignia for formations even down to the company level. Following the French example, officers initially wore metal unit insignia suspended from pocket hangers over their right breast button; enlisted personnel wore cloth versions on the left shoulder. By the 1960s pocket hangers had been phased out and all ranks wore shoulder insignia. Parachute wings were worn above the right shirt pocket; foreign qualification badges went over the left pocket. FAR paratroopers wore a red beret with a silver metal winged dagger copied from the French airborne troops badge; in the late 1960s this was slightly modified with a Laotian-style trident replacing the dagger.

Most FAR equipment came from the United States, whether via the French or directly through one of the US assistance groups. Steel helmets were the basic M1 version. Web gear included the M1945 and M1956 patterns. Early FAR units used the Thompson sub-machine gun, the Browning Automatic Rifle, and French bolt-action MAS 36 rifles. Airborne units took delivery of the M1 Garand in late 1959, followed by the M1 carbine the following year. In 1969 secret deliveries of the M-16 arrived in Laos, and were given initially only to the Palace Guard and paratroopers; standardisation to the M-16 was completed by 1971.

The Royal Laotian Air Force

On 9 September 1954 the French turned over five MS 500 light aircraft to form the Aviation Nationale Laotienne. By 1959 its handful of C-47 and L-20 transports were sufficient to reinforce and resupply the beleaguered government forces during the Sam Neua crisis.

In December 1960, while Kong Le's forces occupied Vientiane, 12 ANL students were brought to Thailand for bombing and gunnery training in T-6 aircraft. This small cadre returned to Vientiane with four T-6 bombers in January 1961 to form the core of the new Royal Laotian Air Force. Thrown immediately into battle, the T-6 bombers were used primarily to demonstrate the government presence in the contested countryside. Pathet Lao gunners quickly found their mark, however, and killed five of the first six RLAF pilots.

The RLAF experienced its next major expansion when the US turned over three T-28 fighter

Staged photograph of Pathet Lao in combat. The uncharacteristically well-equipped force has AK-47 rifles, a B-40 rocket, and a radio operator with what appears to be a Soviet R-105 transmitter. Headgear is a mixture of soft round caps and khaki field caps.

Propaganda shot of Pathet Lao regulars conversing with the local population. Their reed sun helmets are of Chinese origin and were rarely seen among the Pathet Lao.

bombers in August 1963. Over the next four years the RLAF controlled over 30 T-28s, reaching a peak sortie rate of 1,014 combat missions in January 1966.

In October 1966 the RLAF was involved in a failed coup attempt, leading to the exile of several of its most experienced pilots. After training new pilots to replace those in exile, RLAF sorties again began to increase; however, the RLAF remained a fragmented force lacking motivation and effective leadership. In 1968 each RLAF base was consolidated under one commander, eliminating the earlier system whereby each aircraft type was under a separate RLAF command. During the following year, with a pressing need to co-ordinate air and ground assets in Laos, plans were developed to link the Joint Operations Centres in each of the Military Regions to a Combined Operations Centre in Vientiane. This was accomplished by May 1970, resulting in an improvement in RLAF effectiveness. By that time composite wings had been raised at Luang Prabang,

Vientiane, Savannakhet, and Pakse. Each wing consisted of T-28 ground attack aircraft, AC-47 gunships, C-47 transports, H-34 helicopters, and L-19 light planes. A T-28 detachment also operated out of Long Tieng under the control of Gen. Vang Pao. Most RLAF training was conducted under US auspices at Udorn Air Base in Thailand, although a small RLAF training school existed at Savannakhet.

On 20 August 1973 Gen. Thao Ma, the exiled commander of the RLAF who had fled the country in 1966, crossed from Thailand into Vientiane with an armed force. He took control of a T-28, and began to attack the capital while his rebel forces attempted to capture key points in the city. The plane was shot down by ground fire, and the general executed; the rest of the coup force was arrested or fled back into Thailand. Following the coup attempt the RLAF was largely prohibited from action for the final two years of its existence. With fuel and ammunition rationing in effect, T-28 pilots averaged only two hours of flying time a month. In 1975, with the Pathet Lao on the verge of consolidating control, most of the pilots boarded aircraft and headed out on a one-way journey to Thailand.

The RLAF dress and work uniforms were standard FAR issue until 1964. Pilot's wings, worn over the left breast, were the same as the US Air Force version; for a time, US Air Force rank insignia were also worn.

In 1964 distinctive light blue and blue-grey RLAF work uniforms were introduced. A dark blue dress uniform was also issued to officers. A matching dark blue peaked cap was adopted, at first with the FAR cap device, but after 1967 with a distinctive RLAF silver cap badge. Air force personnel also wore a blue overseas cap. A metal RLAF pilot's badge was created in the mid-1960s, in two classes. It was identical to the US Air Force wings, except for the letters 'RLAF' stamped across the top of the central crest. A white version on a blue cloth background was worn on flight suits. On combat missions RLAF pilots relied on an inconsistent US supply system supplemented by items purchased during training in Thailand. Attempts were occasionally made at standardistion; the 1971 T-28 class, for example, was issued flightsuits in a commercial spotted camouflage pattern. More often, standard olive drab suits were worn, usually with a mesh US Air Force survival vest. Squadron insignia went over the right breast.

RLAF rank insignia were identical to the FAR system except that officers' shoulderboards were blue. On missions, pilots often wore subdued rank insignia on their right collar. RLAF service insignia—a winged lotus leaf rosette—went on the right collar.

Government Allied Forces

Laotian Guerrilla Forces

After 1954 Laos inherited a wide variety of irregular militia units which operated in rural villages. Most were misleadingly called 'commandos', numbering some 25 companies in 1955. The commandos were dissolved in 1958, and only a scattering of Auto Defense de Choc militia units remained at the village level. After Kong Le's 1960 coup an expanded Auto Defense de Choc concept was quickly implemented in north-eastern Laos by Lt.Col. Vang Pao. By mid-1961 Vang Pao's forces has swelled into Auto Defense de Choc companies of 100 men each. Selected members from the companies were then sent to Thailand for training the following year. Upon their return they were organised into Special Operations Teams, each equipped with a radio and sent to mobilise militias in surrounding villages.

Also in 1962, entire battalions of Vang Pao's Hmong tribesmen were trained as Special Guerrilla Units. By 1965 the SGUs were bearing the brunt of the fighting for the government as they were helicoptered across the mountainous terrain of north-eastern Laos.

Hmong guerrilla officers from Groupements Mobiles 22 and 23 carry a NVA rocket launcher captured during Operation 'About Face', 1969. They wear OG 107 fatigues.

In the south, the guerrilla programme started in 1961 with a plan to organise the Lao Theung tribes on the Bolovens Plateau and sweep away the Pathet Lao presence. The programme met with initial success, but was cancelled in June 1962. Southern guerrilla units were not raised in force again until 1965. Under Project 'Hark', teams ranging in size from five to over 20 men were used to collect intelligence and conduct small harrassing raids along the Ho Chi Minh Trail.

In 1967 Vang Pao's guerrillas in the north were divided into two groups. Territorial SGU battalions were raised and assigned to a specific location; the original SGU battalions formed in 1962 were renamed Bataillons Guerrillas and gathered into regimental-sized irregular Groupements Mobiles. In 1969 Vang Pao used his three GMs on the Plaine des Jarres to great effect, sparking the need for additional guerrilla regiments.

Taking their lead from the Hmong guerrilla programme, Savannakhet started to organise its first SGU battalions in 1967. These battalions performed admirably during the 1969 capture of Muong Phine, leading to the creation in 1970 of three 500-man Mobiles for country-wide reinforcement duty. The Mobiles were disbanded later in the year and replaced by irregular Groupement Mobiles each of four guerrilla battalions. Because of the larger population base and lack of direct

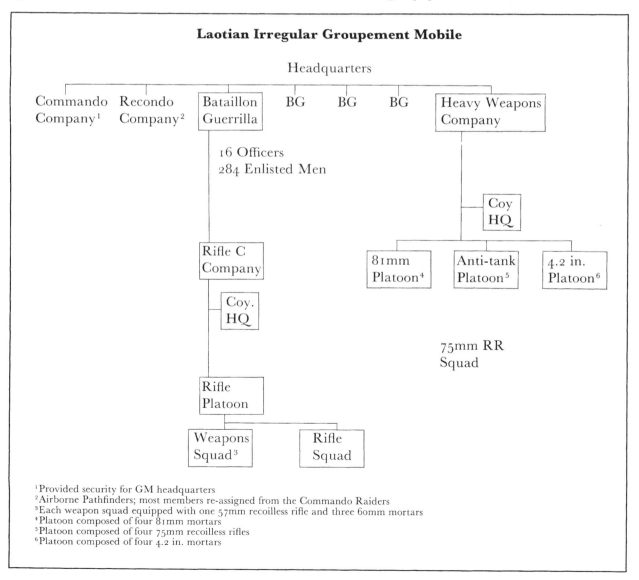

Laotian Irregular Groupement Mobile

Headquarters

Commando Company[1] Recondo Company[2] Bataillon Guerrilla BG BG BG Heavy Weapons Company

Bataillon Guerrilla: 16 Officers, 284 Enlisted Men

Rifle C Company — Coy. HQ — Rifle Platoon — Weapons Squad[3], Rifle Squad

Heavy Weapons Company — Coy HQ — 81mm Platoon[4], Anti-tank Platoon[5], 4.2 in. Platoon[6]

75mm RR Squad

[1] Provided security for GM headquarters
[2] Airborne Pathfinders; most members re-assigned from the Commando Raiders
[3] Each weapon squad equipped with one 57mm recoilless rifle and three 60mm mortars
[4] Platoon composed of four 81mm mortars
[5] Platoon composed of four 75mm recoilless rifles
[6] Platoon composed of four 4.2 in. mortars

Hmong guerrilla and Project 404 attaché climb across a PT-76 amphibious tank captured from the NVA during Operation 'About Face'. While lightly armoured and fitted with a relatively weak 76mm gun, the PT-76 effectively spearheaded all NVA ground offensives across the Plaine des Jarres. (Courtesy Col. Bill Keeler)

enemy pressure on their own military region, the Savannakhet GMs were used extensively over the next three years as reinforcements in battles on the Plaine des Jarres and Bolovens Plateau.

In 1971 Luang Prabang, Nam Yu, and Pakse all created their own guerrilla GMs. Those in Pakse were used on the Bolovens; the Luang Prabang and Nam Yu GMs shuttled between their own regions and reinforcement operations on the Plaine des Jarres.

As stipulated in the 1973 Paris agreements on Laos, the guerrilla forces were scheduled for integration into the FAR. Most of the guerrillas felt unwelcome in the regular army. In addition, decreased wages and other privileges sharply dulled the cutting edge of what had been an effective fighting force, and left them incapable of halting the Communist takeover of the country.

Uniforms and equipment of the Laotian guerrillas reflected a generous, if somewhat erratic, US supply network. The basic uniform initially consisted of OG 107 fatigues; US jungle fatigues were introduced in the early 1970s. At the same time, Vietnam-style leaf camouflage, tiger stripes, and commercial spotted camouflage fatigues appeared in substantial numbers. Other guerrillas insisted on wearing civilian clothes. Little effort was made at uniform standardisation. One exception was the Commando Raiders, an élite airborne strike force within the Savannakhet and Long Tieng guerrilla programmes. Those raiders in Savannakhet wore distinctive dark blue fatigues with a black or blue beret. Enemy uniforms were used on cross-border operations.

Black leather boots were used by the guerrillas throughout the 1960s. By 1971 jungle boots were also worn—a prized piece of equipment not issued to the FAR. Also seen were indigenous olive Bata boots.

Guerrilla headgear was diverse; red FAR berets, baseball caps, patrol caps, US Marine-style utility caps, and civilian hats all found their way on to guerrilla heads. Some units attempted to adopt a green beret with little success. US M1 steel helmets were seen—rarely—in this field.

Nametags were occasionally worn on uniforms, both in black-on-white and subdued versions. Because guerrilla recruits included ex-members of the FAR, rank insignia from the regular army were worn on occasion. A slightly modified system of guerrilla rank insignia was developed in 1971, and seen occasionally on guerrilla officers in the field. Worn on the left collar, the system consisted of an appropriate number of black stars on an olive drab cloth background. No guerrilla branch insignia existed.

Like the FAR, the guerrilla forces spawned a large number of unit insignia. Battalion and GM insignia were developed for the left shoulder, but rarely seen in the field. Occasionally worn were silver metal SGU 'wings' made in three classes, consisting of US-style parachute wings with a Laotian *erawan*—the three elephants on a

An RLAF major in US Air Force flight jacket and cap poses beside a Hmong bodyguard while touring the Plaine des Jarres, 1969. The boy is equipped with an M2 carbine, M26 fragmentation grenades, and US OG 107 fatigues—size small! (Courtesy Col. Bill Keeler)

Order of Battle, Laotian Irregular Forces, Spring 1972

Military Region One (North-west Laos)

Unit	Staging base
Groupement Mobile 11	Luang Prabang
GM 12	Luang Prabang
GM 15[1]	Nam Yu

Military Region Two (North-eastern Laos)

GM 21	Long Tieng
GM 22	Long Tieng
GM 23	Long Tieng
GM 24	Long Tieng
GM 25[2]	Muong Moc
GM 26	Long Tieng
GM 27[2]	Bouamlong
GM 28	Long Tieng
Commando Raider Company (Airborne)	Pha Khao

Military Region Three (Upper Panhandle)

GM 30	Savannakhet
GM 31	Savannakhet
GM 32	Savannakhet
GM 33	Savannakhet
GM 34[3]	Savannakhet
Commando Raider Company (Airborne)	Savannakhet

Military Region Four (Lower Panhandle)

GM 41	Pakse
GM 42	Pakse
GM 43	Pakse

Standard Groupements Mobiles in Military Regions One and Two composed of three battalions. Standard Groupements Mobiles in Military region Three and Four composed of four battalions.

[1] Composite regiment composed of Lao Theung, Lao, and Mien units.
[2] Territorial static defence unit of two battalions.
[3] Reserve training unit of two battalions.

pedestal—replacing the parachute. This was not a parachute wing, but rather was awarded for meritorious service in the guerrilla forces.

The Laotian guerrillas were issued a full range of US equipment, usually of higher quality than that afforded the FAR. During the early to mid-1960s, some pieces of equipment, such as the M-16 magazine pouch, were specially made for use in Laos. By the late 1960s, regular issue US equipment was commonplace. The standard guerrilla web gear was a mixture of the M1945 and M1956 systems. Guerrilla weaponry came from US sources. After 1969 Vang Pao's irregular forces began to receive the M-16 rifle, and standardisation to the M-16 among guerrilla units was completed by 1971. Heavier weapons issued on the battalion level included the M-79 grenade launcher, Browning Automatic Rifle, M-60 machine gun, 60mm mortar, 57mm recoilless rifle, and LAW anti-tank rocket. In 1971 a heavy weapons company was added to each of the Savannakhet GMs. These companies were provided with 75mm

Hmong guerrillas inspect Chinese BJ212 light vehicles captured from the NVA during Operation 'About Face'. (Courtesy Col. Bill Keeler)

recoilless rifles, 81mm mortars, and 4.2in. mortars. Communication gear used by the guerrillas consisted mainly of the PRC-25, HT-1, and HT-2 radio.

Thai Volunteer Forces

Sharing a similar ethnic and linguistic past, Thailand quickly formed a close relationship with the young Royal Laotian government. This led to early Thai military assistance to the fledgling Laotian army, including the loan in 1955 of two H-19 helicopters with contract crews. In 1959 assistance was expanded to include refresher training for the ANL's two parachute battalions.

Thailand watched with great concern Kong Le's successful 1960 coup and the subsequent expansion of Pathet Lao control over Sam Neua province and the Plaine des Jarres. Identifying Vang Pao's guerrillas as being the most effective opposition to

FAR officers tour the Plaine des Jarres in late 1969. The officer on the left wears a red beret with silver FAR airborne badge. Note the enormous stone jars for which the 'PDJ' is named. (Courtesy Col. Bill Keeler)

the Pathet Lao, the Thais sent some 99 members of their élite Police Aerial Reinforcement Unit in early 1961 to help train and organise the budding Hmong guerrilla army. In this rôle the PARU earned a reputation as one of the most effective unconventional warfare groups in South-East Asia.

Over the next nine years Thai assistance was largely limited to training and occasional artillery support. One exception was the intervention in March 1968 by an élite infantry unit airlifted into a besieged mountaintop radar facility in northern Laos. The unit was quickly withdrawn after evacuating key personnel. Thai air support began in April 1964 when volunteer T-28 pilots known as the B Team, began flying in northern Laos. They were withdrawn in 1972 after providing 40 per cent of the total T-28 sortie rate.

In February 1970 the situation in northern Laos

deteriorated dramatically when Vang Pao's headquarters at Long Tieng came in danger of being overrun. The Thais flew in artillery and security troops to defend the base, freeing Hmong forces for operations behind NVA lines. The following year more Thai regular army units from the 8th Regiment were introduced for operations north and east of Long Tieng base.

Because the Hmong population was suffering heavy casualties and could not continue to contain the repeated NVA annual offensives alone, plans were developed in early 1971 to commit a larger Thai force to Laos. Under Project 'Unity' volunteer infantry and artillery battalions were assembled for use on the Plaine des Jarres and Bolovens. The first two 'Unity' battalions were fielded on the northern Bolovens in April 1971 with spectacular results. The 39th NVA Regiment, mistaking the Thais for a typically ineffective FAR unit, attacked the positions by night. Next morning 143 Vietnamese bodies were counted on the outer perimeter; no Thais were killed.

With confidence high due to the Thai performance on the Bolovens, three further battalions were sent north to participate in Vang Pao's successful capture of the Plaine des Jarres in the summer of 1971. Once on the plain Thai 105mm and 155mm artillery bases were dug in as the backbone of Vang Pao's defence strategy. In December 1971 Thai units on both the Plaine des Jarres and the Bolovens Plateau suffered heavy casualties during simultaneous NVA offensives. Further battalions were raised and sent to reinforce besieged Long Tieng and Pakse during early 1972. A further five 'Unity' battalions were garrisoned at Xieng Lom in north-west Laos to help secure the Lao-Thai border. During the summer of 1972 some 25 Thai battalions, loosely grouped into five mobile groups, were operating in Laos. They provided vital reinforcements for the beleaguered Laotian irregulars, without which Laos would probably have fallen by 1971.

In 1973, with the ceasefire in effect, volunteer Thai battalions maintained defensive positions on the Bolovens, the Plaine des Jarres, and the strategic crossroads along Route 13 north of Vientiane. The last 'Unity' battalions were withdrawn in June 1974.

The Thai volunteers were equipped much like the Laotian guerrillas. The most common uniform was the OG 107 shirt and pants used by the Royal Thai Army; it was not unusual, however, to see a wide range of other fatigue variations, and camouflage. One Thai unit was even supplied with surplus stocks of South Vietnamese Police Field Force leopard-spot fatigues, which gives some indication of the varied supply network that fed the 'Unity' programme. Headgear was a mixture of jungle hats, baseball caps, and Royal Thai Army patrol caps. Some Thai battalions with cadres drawn from the Special Warfare Centre at Lopburi were given red, black, or blue berets to denote their élite status. Included in this category was the 703 Reconnaissance Unit, a 50-man commando force which arrived at Long Tieng in the spring of 1972 and returned after only two short months of action. Members of this unit wore a blue beret during their brief deployment.

Because of the unpublicised nature of the Thai

Reinforcements from the FAR 101ᵉ Bataillon de Parachutistes arrive on the Plaine des Jarres, late 1969. The unit designation has been spelled on the ground in shell casings to prevent accidental bombing. (Courtesy Col. Bill Keeler)

involvement, no nametags, rank, or branch insignia were worn. Many battalions did evolve their own unit insignia, including beret badges and shoulder insignia. More common were metal 'beercan' devices mounted on cigarette lighters. The symbol for Project 'Unity', a pair of temple dragons below a Laotian *erawan*, was used on flags and shoulder insignia, but rarely seen in the field.

Military equipment for the 'Unity' battalions was drawn from US sources. The MI steel helmet, M1956 web gear, cotton magazine bandoliers, and the Lightweight Tropical Rucksack were all used. The standard rifle was the M-16, although small numbers of 'Swedish Ks' and M3 .45 calibre sub-

machine guns were seen. Also issued to every battalion was the M-79 grenade launcher, 81mm mortar, M-60 machine gun, and LAW rocket. Heavy Weapons Companies assigned to every battalion had the 90mm recoilless rifle and 4.2 in. mortar. Communications gear used by Thai forward air guides included the PRC-77 and HT-2 radio.

US Military Assistance Groups

US military involvement in Laos was characterised by continuous diplomatic restraint. Abiding by the 1954 Geneva Agreement on Indochina, the US was forbidden from opening a Military Assistance and Advisory Group in Laos. Instead, a civilian alternative called the Program Evaluation Office was started in December 1955 in Vientiane. For four years the PEO primarily handled equipment deliveries to the ANL, while allowing the 'Mission Militaire Française près du Gouvernement Royale

Three versions of the US HT-2 hand-held transceiver, used by guerrilla and Thai volunteer forces in Laos to co-ordinate artillery and close-air support. The set on the right is used with a base station. A self-destruct button on the lower right of the left-hand model sends an electrical surge through the set, rendering it useless in the event of imminent capture. (Courtesy Glen Thibodeaux)

1: Lt. Deuane, 2^{e} Bataillon Para., 1960
2: Guerrilla, Auto Defense de Choc, 1961
3: Capt., US 'White Star' MTT, 1961-62

A

1: Sgt. Directorate of National Co-Ordination, 1962-64
2: Maj., Police Aerial Reinforcement Unit, 1961-64
3: Maj. Gen., Forces Armées Royales, 1961-75

B

1: Pathet Lao officer, 1975
2: Pathet Lao guerrilla, 1968
3: Laotian Commando Raider, 1969

1: Roadwatcher, 1965-68
2: Maj., Savannakhet GM guerrillas, 1972
3: Guerrilla, Luang Prabang forces, 1971

D

1: Gen. Vang Pao, 1969
2: Hmong guerrilla, GM 21, 1969
3: Lt., RLAF, 1971

E

1: Thai forward air guide, 1971
2: Thai 'Unity' volunteer, 1972
3: Hmong guerrilla, 1972

2

1

3

F

1: Brig. Gen. Thao Ly, 2nd Div., 1971
2: Capt., US Army Special Forces, Project 404, 1972
3: Maj., 714^{e} Bn. Para., 1974

1

3

2

4

6

5

7

8

Laotian and Thai insignia; see text for identification

H

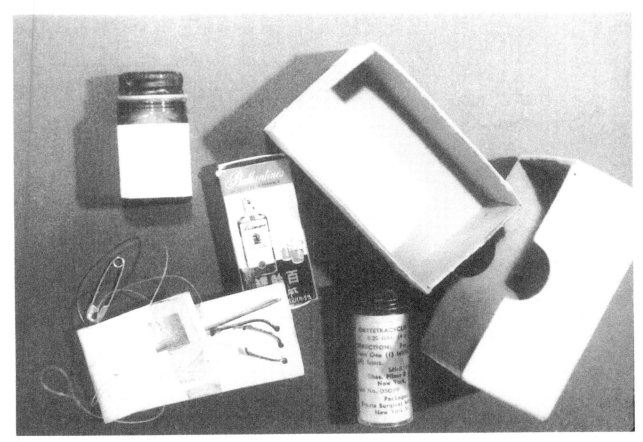

An NVA 'survival kit' found on the body of a sapper killed while attacking Xieng Khouang airfield in February 1970. The kit includes matches, a safety pin, fishing hooks, water purification tablets, and medicine. The last two items—incredibly—are of US manufacture. (Courtesy Col. Bill Keeler)

Laos', the residual French military mission, to conduct training for the Laotian forces.

In 1959, frustrated with the state of the ANL, the PEO entered into meetings with its French counterpart, resulting in an agreement allowing greater US participation in direct Laotian training. By July 1959 the first US Army Special Forces teams, still acting as civilian members of the PEO, arrived in Laos. They began to work closely with the ANL, building training facilities, distributing new equipment, and instructing the Laotians in counter-insurgency techniques.

In August 1960 the PEO was temporarily paralysed by the Kong Le coup. By the next month, however, when it became apparent that the naive paratroop captain was handing too much to the Pathet Lao, the PEO began to assist the counter-revolutionary forces massing at Savannakhet. In December, Special Forces teams were actively involved in co-ordinating the retaking of Vientiane.

On 19 April 1961 the PEO was renamed the US Military Advisory and Assistance Group, Laos.

The field training teams were renamed 'White Star' Mobile Training Teams and each was assigned to a specific FAR battalion, following the FAR into the field and advising them on combat operations. For example, the Special Forces members assigned to the FAR 55ᵉ Bataillon de Parachutistes made a combat jump with their Laotian counterparts at Nam Tha on 27 March 1962. Other 'White Star' units were used to raise unconventional forces among the Lao Theung tribesmen on the Bolovens Plateau and the Hmong hill-tribes in the north.

In August 1962, in accordance with a new set of Geneva Agreements, the US MAAG was closed and the 'White Star' teams evacuated through designated exit points. To handle military procurement for the FAR a Requirements Office was attached to the US Embassy. A small military

Gen. Vang Pao, left, the famous guerrilla commander of
northern Laos, attends a funeral for a Hmong T-28 pilot killed
in action, 1970. Two guerrillas in the rear wear red berets with
gold FAR badges. Note the FAR parachutist wings worn by the
guerrilla in the centre. (Courtesy Col. Bill Keeler)

attaché's office also remained in Vientiane. By
early 1964, with the Pathet Lao massing for an
offensive in the north, the small US military
presence was deemed inadequate. Approval had
already been given for Project 'Waterpump', a
covert programme tasked with improving and
expanding the Lao Air Force. Composed of US Air
Force Air Commandos, 'Waterpump' arrived in
Thailand in March 1964, and was providing vital
assistance to the RLAF by the start of the Pathet
Lao offensive in May.

Initially, diplomatic restraints curtailed the
activities of Project 'Waterpump'. Forbidden from
being stationed in Laos, personnel were forced to
shuttle back to Thailand at night. By day the Air
Commandos were able to train Laotian pilots,
organise an augmentation force of Thai T-28
volunteers, and set up a string of Air Operations
Centers across Laos. A small number of combat

missions were also flown by 'Waterpump' pilots.
The number of 'Waterpump' personnel working in
Laos grew throughout 1965.

In early 1966 the US Joint Chiefs of Staff
approved a plan to organise an augmentation
group in Laos that would provide a more
permanent operating framework for the US
military personnel then assigned to Laos on
temporary duty. In October the 117 US servicemen working in Laos were designated assistant
attachés and assigned to a new umbrella organisation named Project 404. Support for Project 404
was channelled through the Deputy Chief, Joint
US Military Assistance Group, Thailand. In
practice, Project 404 remained tightly controlled
by the US Ambassador in Vientiane.

Much of Project 404 was geared toward assisting
the RLAF. US Air Force teams operating at each
of the major RLAF bases assisted with mainten-
ance, training, and communications. Other 404
attachés worked with the FAR, though never to
the intimate degree of their 'White Star' predeces-
sors. In 1971 Project 404 took on the added mission
of completely retraining and reorganising the FAR

along US lines. A major rehauling of the FAR was vital because the US envisioned an eventual military withdrawal from South-East Asia, meaning that guerrilla forces would be scaled down and the FAR would once again have to assume full responsibility for national security. Accordingly, US Army Special Forces teams assigned to 404 helped to set up three regional training centres and retrain all FAR infantry and artillery units. Other 404 personnel were used on strategic reconnaissance and sensor-planting operations in Laos until March 1973.

Also in 1971, the need arose to train a large number of Laotian, Hmong, and Thai forward air guides for the irregular forces fighting on the front lines. Responsibility for this training fell on Project 'Waterpump', which since 1966 had reverted to a small US Air Force training detachment at Udorn Air Base, Thailand. US Air Force Combat Controllers assigned to 'Waterpump' trained the indigenous guides both in Thailand and Laos.

In 1973 the Deputy Chief, JUSMAG Thailand,

moved to Vientiane and was renamed the Defense Attaché, Laos. In line with the general US disengagement from South-East Asia, the Defense Attaché was instructed to reduce the 180 military servicemen in Laos down to 30 persons by December 1974. Accordingly, Project 404 was shut down in June 1974, leaving four dozen US military personnel in Laos on the eve of the country's capitulation.

The original PEO personnel in Laos wore uniforms that reflected the civilian status of their organisations. Until 1959 this was not misleading, since the PEO was composed of civilians and predominantly involved with equipment procurement. When the first Special Forces teams arrived

Three RLAF pilots pose at Savannakhet airbase, 1971. The pilot on left wears commercial spotted camouflage and a US M1956 pistol belt with strobe light attached. The centre pilot wears a black flightsuit with US Air Force mesh survival vest, and cloth RLAF pilot's wings over his left breast. The officer on the right wears US leaf-pattern camouflage with mesh survival vest. The Laotian national emblem with 'Laos' tab, worn while training in Thailand, is on his right shoulder. All wear red 'T-28 Fighter Bomber' caps issued to their graduating class. See Plate E3. (Courtesy Hap Lutz)

in July 1959, many continued wearing civilian clothes. Others wore nondescript fatigues devoid of US military insignia.

When the 'White Star' teams became overt in the spring of 1961 they were officially allowed to wear US military uniforms, including their green berets. A 'White Star' beret flash consisting of yellow crossed arrows and a dagger on a red background was developed, but not worn due to continued diplomatic sensitivity. In the field a variety of headgear was worn, including a locally-produced French-style bush hat in camouflage material. While on operations, 'White Star' teams were equipped with M1 or M2 carbines, M3 sub-

machine guns, Bren light machine guns, and occasionally Sten or Owen sub-machine guns.

Project 'Waterpump' crews initially wore civilian clothes during their first years of deployment, making it common to see Americans in shorts, Hawaiian shirts, and shower sandals working with the RLAF at Wattay airfield outside Vientiane. Once in the field, members of 'Waterpump' adopted military clothing of their choice, usually including the trademark Air Commando bush hat. National and unit insignia were not worn on field uniforms.

By 1971, uniform restrictions had been lifted on Project 404 members assigned to Laotian training bases, allowing them to wear full US military uniforms and insignia. Combat controllers who escorted indigenous forward air guides back to Laos wore either civilian clothes or camouflage uniforms without identifying insignia.

Thai volunteer forces in Laos, 1971. The soldiers in the foreground wears Thai leaf camouflage with zippered shirt, and an M-59 grenade on his US M1956 pistol belt. In the centre, a volunteer has a cotton M-16 magazine bandolier and four 3.5 in. rockets slung over his shoulders. Note the Buddhist amulets around the neck of the soldier in the rear.

Communist Forces

Pathet Lao Uniforms and Equipment

Pathet Lao uniforms and equipment showed a mix of Chinese and North Vietnamese influence. Two basic uniforms existed. The first consisted of a khaki shirt and trousers delivered through NVA channels. This was worn with either a matching Mao-style soft cap with black leather peak; a more bulbous soft cap like that worn by the Burmese Communist Party; or a khaki cloth field cap. The second uniform, in predominent use by the early 1970s and still worn by the current Communist government forces, was a green shirt and trouser combination with matching soft Mao cap. Shirts came in several styles, the most common being a two-pocket version with rectangular pocket flaps, single exposed buttons, and shoulder straps; buttons were of green plastic. A second variation had four shirt pockets and pointed pocket flaps, while a third style had neither shoulder straps nor pockets. Pants were generally cut in the NVA style, with two front pockets, one rear pocket, and simple straps for gathering the pant's leg at the ankles. A second, more spartan variation was less baggy and had no ankle-straps. Footwear was either the NVA canvas combat shoes or rubber sandals. Belts were of olive drab or khaki canvas with plain silver buckles. Black leather belts were also issued to officers.

As with uniforms, the Pathet Lao used Chinese and North Vietnamese equipment, channelled primarily through the NVA. A full range of ChiCom and NVA ammunition pouches, equipment belts, grenade pouches, and canteens were used; a canvas or leather map case was commonly worn by officers as a symbol of rank. The Pathet Lao wore no branch, unit, or rank insignia during the war. Since 1975 a round metal cap device bearing the flag of the People's Democratic Republic has been adopted.

The North Vietnamese Army

Despite some instances of blatant intervention by NVA Main Force regulars before 1968, Hanoi attempted until 1967 to portray the fighting in Laos as a civil war. During that time the NVA created two units that handled most of its missions

Hmong guerrillas consult a map during operations around the 'PDJ', 1971. They carry M-16 rifles and the LAW rocket.

inside Laos. The first, known as Doan 559, was an umbrella group created in May 1959 to secretly transport men and military equipment to South Vietnam via the Laotian panhandle. The second unit was Doan 959, created in September 1959 to provide military and political assistance to the Pathet Lao. Doan 959 had a forward headquarters located in Sam Neua, and co-ordinated NVA advisory and combat activity in Laos until it was de-activated in 1973.

The NVA attempt to disguise their presence in Laos before 1968 was reflected in their uniforms. Advisors to Pathet Lao battalions were instructed to adopt the complete uniform and equipment of the Laotian Communists. Main Force elements directly confronting government forces also adopted Pathet Lao dress. A Vietnamese prisoner captured in 1966 after an attack on the Dong Hene Military Academy, for example, was found wearing 'a light khaki uniform, light khaki canvas shoes, and a light khaki loose top peaked hat', all of which were associated with the Pathet Lao. By 1968, however, with the NVA presence an open secret, complete NVA uniforms with rank insignia were openly worn on the Laotian battlefield.

Order of Battle, North Vietnamese Army in Laos, 1971–2

Unit	Location, 1971	Location, 1972
312 Div.[1]	vic. PDJ	vic. PDJ[2]
316 Div.[3]	vic. PDJ	vic. PDJ[4]
968 Div.[5]	Khong Sedone, Saravane, Paksong	Route 23, Saravane, Paksong
1 Regt.[6]	Muong Nong (Ho Chi Minh Trail)	?
8 Regt.	—	vic. Luang Prabang
29 Regt.	?	vic. Muong Phalane
41 Regt.[6]	?	vic. Tchepone[7]
49 Regt.[6]	?	vic. Tchepone[7]
101 Regt.	?	eastern Bolovens[8]
335 Independent Regt.	vic. Luang Prabang[9]	Bouamlong
766 Independent Regt.	vic. PDJ	vic. PDJ
866 Independent Regt.	vic. PDJ	vic. PDJ

Excludes NVA units assigned to Doan 559 not directly in conflict with Laotian government forces. NVA divisions include reconnaissance, sapper, artillery and armoured elements during dry season offensives.

[1]Subordinate units include 141, 165 and 209 Regiments.
[2]Withdrawn in March 1972 for South Vietnam operations.
[3]Subordinate units include 148, 174 and 176 Regiments.
[4]Bulk of division withdrawn in early 1973.
[5]Subordinate units include 9, 19 and 39 Regiments.
[6]Operating under Doan 559.
[7]Operating in vicinity of Muong Phalane by early 1973.
[8]Committed south of Saravane in early 1973.
[9]Decimated by government guerrilla forces and withdrawn to North Vietnam for refitting.

The Plates

A1: Lieutenant Deuane, 2ᵉ Bataillon de Parachutistes, 1960

On 9 August 1960 Capt. Kong Le led his 2ᵉ BP in taking over Vientiane. Lt. Deuane, a company commander, emerged as Kong Le's chief rival; and in 1964 he broke ranks and formed a pro-Communist faction called the Deuanist Neutralists, noted for an attack on the Sala Phou Khoun road junction in 1972. After the fall of Laos in 1975 its 800 armed members were absorbed into the Pathet Lao.

In 1960 the Laotian paratroopers were the most pro-French soldiers in the ALN, a fact reflected in their uniforms and equipment. Lt. Deuane, photographed on the day after the coup, wore a French airborne camouflage 1947/52 smock and 1947/51 trousers and a small bush hat in French camouflage material, as issued to his company. On his US pistol belt is a 9mm Luger, a French OF 37 grenade, a first aid pouch, ammunition pouch and bayonet; he carries the US M1A1 carbine, issued in small numbers to the 2ᵉ BP. On his right jacket pocket are first lieutenant's rank insignia pinned on a red cloth rectangle. Noticeably absent are paratroopers' jump wings.

A2: Guerrilla, Auto Defense de Choc, 1961

This guerrilla, based on a photograph, is relatively well equipped for the time. Both his olive drab jacket and faded camouflage trousers are French paratrooper issue. His leather boots, M1936 cartridge belt and M1 carbine are American. The red beret, worn by the entire FAR, is a personal acquisition and was never adopted by either the ADCs or subsequent irregular units. The white cloth tied to his upper sleeve is a unit identification.

A3: Captain, US 'White Star' Mobile Training Team, 1961–62

In April 1961 US Army Special Forces training teams operating under civilian cover were allowed to wear military uniforms as part of an overt Military Advisory and Assistance Group. This captain is assigned to a Mobile Training Team operating with a FAR battalion for six months. He wears Laotian parachute wings over the right breast pocket of his OG 107 fatigues, indicating time spent with one of the three airborne battalions from FAR Groupement Mobile 15 at Seno. As trainers, 'White Star' teams regularly went into the field with their units. The captured Soviet PPSh-41 sub-machine gun with 71-round drum is evidence of the Pathet Lao threat they faced.

B1: Sergeant, Directorate of National Co-ordination, 1962–64

The December 1960 drive to oust Kong Le from Vientiane was spearheaded by Groupement Mobile Speciale 1 under command of Lt.Col. Siho Lamphouthacoul. After consolidating his hold over the capital Siho was rewarded with command of a new paramilitary police organisation, the DNC. It combined both intelligence and commando missions, but was tarnished by a growing reputation for involvement in corrupt activities. In April 1964, Gen. Siho used the DNC to take over the capital in a successful coup d'état. While expanded to include a crack airborne regiment, the DNC saw little combat and was used mainly as Siho's personal guard. In February 1965 a countercoup headed by Gen. Kouprasith Abhay succeeded in driving Gen. Siho into exile before the DNC could be brought into action. As a result the unit was quickly dissolved, with its paratroop elements forming the core of the new Groupement Mobile 21 Aéroportée.

DNC paratroopers wore dark blue fatigues to distinguish them from the rest of the Laotian armed forces. The tiger insignia worn on the left shoulder, bearing the Laotian inscription 'Revolutionary Forces', commemorates Siho's involvement in the recapture of Vientiane in 1960. A black beret is worn American style, reflecting the training DNC cadres received from the Thai PARU in 1962. A distinctive set of gold metal parachute wings in three classes were distributed for DNC use in 1964.

Guerrilla officers from GM 21 discuss operations, 1971. The soldier on the right wears M1945 suspenders, and an OG 107 shirt with an extra pocket sewn on the upper sleeve.

Because of Siho's expanded influence in Vientiane, the DNC was the best-equipped unit in Laos. Reflecting their privileged status, they were the first unit to be completely outfitted with the M2 carbine.

B2: Major, Police Aerial Reinforcement Unit, 1961–1964

The PARU was formed in 1951 as an unconventional warfare unit to operate behind enemy lines in the event of a Chinese invasion of South-East Asia. When no such invasion materialised by the mid-1950s, the PARU was maintained as an élite outfit that could conduct covert operations in Thailand and across the border. In 1961 its first major assignment was the training of Hmong guerrillas loyal to Vang Pao. Close to one hundred PARU commandos, divided into 13 teams, were operating in northern Laos by July 1961. In later

A Thai reconnaissance trooper arrives at Long Tieng headquarters during the heavy NVA siege in early 1972. M-79 grenade pouches are slung over each shoulder.

Oudone Sananikone, FAR Chief of Staff, in 1971. The jacket, made in Vientiane, is a copy of French-made dress khakis worn by the ANL General Staff as early as 1954. The peaked cap and buttons bear the FAR wreathed trident. General Headquarters insignia are worn on both collars. Gen. Oudone, the original commander of the 1st FAR Armoured Reconnaissance Squadron, retains his unit insignia on the left shoulder.

C1: Pathet Lao officer, 1975
By 1969 the Pathet Lao claimed to field 45,000 regulars in 110 battalions. This number had dropped slightly to 39,000 in April 1971, of which 25,000 were considered main force elements. While impressive on paper, the Pathet Lao fought less as time went on, leaving the NVA to conduct all major battles between 1968 and 1973.

An NVA soldier belonging to a *Doan 559* support unit poses along a raised section of the Ho Chi Minh Trail north of Tchepone, in a photograph captured by an allied reconnaissance team, late 1972. **(Courtesy Shelby Stanton)**

years the PARU used its base in Hua Hin, Thailand, to train members of the DNC and the first Laotian SGU battalions. A smaller number of PARU were retained at Long Tieng to continue officer training for the Hmong. By the early 1970s, experienced veterans from the PARU provided many of the forward air guides attached to the Thai volunteer battalions and Hmong Groupements Mobiles. In 1973, after 12 years of action, the last PARU teams departed Laos.

PARU members wore Thai copies of OG 107 fatigues, which were standardised throughout the Thai Border Patrol Police, to which the PARU was subordinate. The M1945 suspenders are being worn with the straps crossing the chest. A first aid pouch and .45 calibre holster hang from the pistol belt. The M3 sub-machine gun was issued to PARU teams in Laos, along with the M1 carbine, BAR, 2.34 in. rocket launcher, and 60mm mortar. The PARU wore an American-style black beret, later adopted by the entire Border Patrol Police. Although teams were supposed to go into Laos without identifying insignia, metal Thai police wings are being worn as a beret badge.

B3: Major-General, Forces Armées Royales, 1961–1975
This dress uniform is taken from a photo of Gen.

The dark green shirt and trousers were standard among Pathet Lao regulars by the early 1970s. Headgear is the soft dark green Mao cap with black leather visor and strap with gold buttons. Privileged officers wore the NVA combat shoes with rubber soles; more common in the lower echelon were rubber sandals. Accoutrements were from both Chinese and Vietnamese sources. Often seen among officers was the Chicom brown leather map case worn across the shoulder. Ballpoint pens displayed prominently in the shirt pocket were an additional status symbol for officers. A Soviet RGD-5 hand grenade hangs from his trouser pocket. He holds a Soviet 7.62mm AK-47 assault rifle with folding stock.

C2: Pathet Lao guerrilla, 1968
While fielding conventional armour, artillery, and air defence units, the Pathet Lao were predominantly a lightly armed guerrilla force. Regional forces and village militia composed slightly less than half of its total force structure. This member of the regional forces, taken from a staged propaganda picture, wears a khaki shirt with two pockets and exposed plastic buttons: supplied by the North Vietnamese to both the Pathet Lao and Viet Cong,

Thai forward air guide, right, pictured near Muong Kassy, early 1973. He holds an HT-2 radio in his right hand, and wears a US Air Force survival vest. A second HT-2 hangs on the post to his right. The FAR officer in the foreground wears commercial spotted camouflage with the shoulder insignia of the FAR Fifth Military Region.

it is commonly called the NVA 'export' uniform. Olive drab pants and khaki field cap were also supplied by the Vietnamese. He is armed with an AK-47 rifle and Soviet 7.62mm Tokarev pistol.

C3: Laotian Commando Raider, 1969
Several commando raider units were raised late in 1968 for specialised 'behind the lines' missions including prisoner-of-war rescue, cross-border raids, reconnaissance, and crash-site recovery. Teams operating out of north-east Laos occasionally wore North Vietnamese uniforms and carried AK-47s. Savannakhet CR teams sometimes used Pathet Lao uniforms when operating along the Ho Chi Minh Trail. Developed for the Commando Raiders in 1969 was the two-shot 2.75in. rocket launcher, modified from the rocket pod on an O-1 spotter aircraft. High explosive heads were fitted, and a simple tripod attached to the launching tubes. Rocket missions extended into North Vietnam and were successful in harassing the NVA

A UH-1C Huey gunship flown by Thai volunteers touches down at Pakse airbase, 1972. It is armed with 2.75 in. rockets and an XM-21 7.62mm flexgun system. A white horse—the radio callsign for UH-1C support in Laos—is painted on the nose of the helicopter.

in what the Communists had considered 'safe' territory. Several versions of the rocket launcher were made, including a six-shot wheeled type.

D1: Roadwatch team member, 1965–1968

In 1965 Laotian roadwatch teams began to operate along the Ho Chi Minh Trail. Initially a crude operation, the roadwatching project was streamlined into an effective intelligence network that extended from the Nape Pass in the northern panhandle down to the Cambodian border. Typically, teams walked to their objective, established a rear operating station, then set up forward monitoring posts overlooking the trail. By rotating the forward units, a resupplied roadwatch team could stay in the field for up to six months.

The roadwatching concept evolved to its highest form in 1967, when improved communications gear allowed the teams to relay their information directly to a processing centre at Nakhon Phanom Air Base in Thailand. The teams could then, in theory, instantly alert air power to targets on the trail. By 1968, however, the roadwatchers had become outdated when improved aircraft sensors and the use of airdropped sensor devices under Project 'Igloo White' delivered more expansive coverage than the teams.

The roadwatch teams wore OG 107 fatigues or commercial spotted camouflage uniforms; head-gear included patrol caps, utility caps, or berets. The M2 carbine was standard; for reinforced teams, 9mm 'Swedish Ks' were also used. Footwear included the Bata boots seen here, leather boots, or sandals. An indigenous rucksack carries enough rations for three days at a forward monitoring post.

Sophisticated equipment was usually not supplied to teams until 1967, when an indigenous counting device was developed for their use. Built from a modified US Air Force survival radio, the counter permitted indigenous personnel to press picture-coded keys as many times as they saw a particular piece of equipment pass by on the trail.

This information would then be gathered by an orbiting aircraft and relayed to Nakhon Phanom. In this way, language difficulties were bypassed. The success of the counting device eventually led to several models being developed.

D2: Major, Savannakhet GM guerrilla forces, 1972
Although the Savannakhet guerrillas were assembled into regiments and used as light conventional infantry, they maintained a flexible structure unlike a regular armed force. Rank and seniority meant little; performance and motivation propelled guerrillas into leadership positions. This major—already a GM commander—carries a .45 calibre M1911 on his M1956 belt as a status symbol. He also wears the modified guerrilla rank insignia on the collar of his shirt. His denim jacket, pants, and boots are all civilian items. The scarf is for identification in the field. Pinned to his left jacket pocket is a guerrilla forces badge, first class, awarded for meritorious service in the irregular army.

D3: Guerrilla, Luang Prabang guerrilla forces, 1971
This guerrilla, taken from a photograph of GM 11 after it mauled the NVA 335th Regiment during the recapture of the king's farm outside Luang Prabang in April 1971, represents one extreme of dress. His hat, shirt, and pants are civilian items; footwear is the US black leather boot. Hanging from his M1956 pistol belt are M-59 and M26 fragmentation grenades; he carries both the M-16 and a LAW anti-tank rocket, distributed only on special missions.

E1: General Vang Pao, guerrilla forces commander, 1969
The most famous anti-Communist guerrilla commander in South-East Asia, Vang Pao began his military career as a commando operating with French forces. In 1954 he participated in Operation 'Condor', the aborted attempt to relieve the besieged garrison at Dien Bien Phu. After the French departed he took a commission in the newly independent ANL. In 1961, as commander of Hmong militia in north-east Laos, Lt.Col. Vang Pao became the most effective government officer operating against the Communist forces in the hills around the Plaine des Jarres. Promoted to Military Region commander in 1965, Gen. Vang Pao

eventually survived nine air crashes and one serious bullet wound while leading his guerrilla forces against as many as three NVA divisions.

Vang Pao preferred to be in the field with his men. He is depicted as seen during a 1969 tour of the Plaine des Jarres after its capture from NVA forces. Reflecting his penchant for US military clothing, he is wearing a personalised US Air Force flight jacket given to him during a tour of Nakhon Phanom air base in 1968. The three silver stars of FAR major-general are sewn onto the shoulder;

Hmong forward air guide, 1972. He wears a US M1956 pistol belt, Universal Pouch, USAF strobe light, and canteen. On his right breast is an FAG qualification patch designed for Lao students passing through training conducted by US Air Force Combat Controllers in Thailand.

the patches and pilot's wings are both USAF insignia. The bush hat, often worn by Hmong officers, was favoured by Vang Pao.

E2: Hmong guerrilla, Groupement Mobile 21, 1969

'About Face', the 1969 operation to recapture the Plaine des Jarres, was the first multi-regimental guerrilla operation in Laos. The mobile tactics of the guerrillas proved highly effective, catching the NVA off-guard and resulting in the capture of millions of dollars worth of Communist hardware. The 3.5in. rocket launcher was used by the guerrillas during the operation. The M2 carbine was still in widespread use, although some battalions had already taken delivery of the M-16. The HT-2 radio slung around his neck was used to communicate between guerrilla forces. An M56 canteen hangs from an M1956 pistol belt. The commercial spotted camouflage pants and cap were procured through South Vietnam.

Joint US-Thai special warfare team in Savannakhet Province, late 1972. The US Special Forces trooper wears a STABO rig and USAF survival vest. He is armed with an M-203, .38 revolver, and shortened M-79 grenade launcher. A smoke grenade, albumin container, and Laotian M-16 magazine pouches hang from his web gear. The Thai soldiers wear indigenous rucksacks. Their M-16 barrels are painted in bright green and brown 'camouflage' to prove their fearlessness in battle. (Courtesy Shelby Stanton)

E3: Lieutenant, Royal Laotian Air Force, 1971

Photographed as a member of the graduating class for T-28 pilots at Savannakhet airbase, Thailand, he wears the embroidered baseball cap given to graduates: it bears his wings above 'T-28/Fighter Bomber'. A good deal of latitude was noted in flight clothing; this officer wears US leaf-camouflage combat fatigues, with his rank insignia sewn to the collars, and the national title and emblem worn on the right shoulder while training in Thailand. Over the fatigues he wears a USAF survival vest, and a pistol belt supports a .45 automatic in a commercial black leather holster.

F1: Thai forward air guide, 1971

During heavy fighting on the Plaine des Jarres and the Bolovens, forward air guides were the link between ground units, fire support bases, and orbiting FAC aircraft. Although Lao nationals were trained later in the war, most FAGs were Thai civilians chosen because of their English language skills. The HT-2 tactical radio allowed for line-of-sight communication with ground units in his vicinity and low orbiting spotter aircraft. The CAR-15 carbine, carried on an improvised assault sling, was favoured by forward air guides. Carried on an M1956 pistol belt is a .38 calibre holster and a

compass case. His faded tiger-stripe fatigues and camouflage hat were privately purchased during R&R in Thailand.

Anti-Communist guerrillas of the United Lao National Liberation Front in formation before departing on operations against the Lao People's Democratic Republic, 1985. Weapons include the RPG-7, AK-47, M-16, and LAW anti-tank weapon.

F2: Thai 'Unity' volunteer, 1972

The Thai volunteers in Laos ranged from seasoned veterans who had fought in South Vietnam to individuals with no prior military experience. In spite of this, they acquitted themselves well in the defensive rôle for several years of heavy fighting. Press reports from Laos describing the unshaven and unprofessional appearance of the volunteers failed to take into consideration that they usually spent their entire tour in Laos on continuous combat operations.

This soldier's M-16 was standard issue. He wears M1945 suspenders, a cotton M-16 magazine bandolier, and a Thai army field cap with Royal Thai Army insignia, evidently a private acquisition since the Thai army did not equip the volunteer forces.

F3: Hmong guerrilla, 1972

Although successful in tying down thousands of NVA soldiers who would have been used in South Vietnam, Vang Pao's guerrilla army took tremendous casualties. By 1971, only a constant supply of reinforcements from Thailand and Savannakhet kept north-eastern Laos from being overrun.

This guerrilla carries the M-16, which was supplied to all Hmong units in late 1970. The jungle fatigues and camouflage jungle hat were received while attending a retraining cycle with his GM in Thailand. M-59 baseball grenades hang from his M1956 pistol belt. At his feet is a PPN-18 Beacon, used in conjunction with the F-111 fighter-bomber during medium altitude, off-set bombing around the Plaine des Jarres.

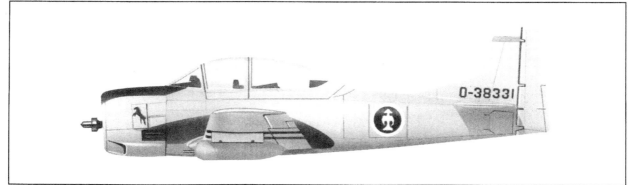

Royal Laotian Air Force T-28A ground-attack aircraft, 1963-75. Originally developed as a trainer, the two-seat T-28A was the primary fighter-bomber of the RLAF. During their service in Laos the T-28s were the first aircraft to strike targets along the Ho Chi Minh Trail, and (flown by 'B Team' volunteers in August 1964) inside North Vietnam. They also took part in two coup attempts. All RLAF T-28s remained in the light grey and black finish in which they were delivered. Tail and nose art was not uncommon; this stallion badge, in blue, was a tribute to the respected Gen. Thao Ma, the RLAF commander who went into exile in 1966—his name means 'horse' in Laotian. Roughly half the T-28 fleet had the red and white national markings painted directly on the fuselage; the rest had Laotian markings painted on removable sliding plates fitted into a three-sided frame on the fuselage side, enabling them to be quickly replaced by USAF, Thai, or no markings at all, depending upon the pilot and mission. T-28 pilots discovered early on that the enemy held their fire until they could see that the aircraft had expended its ordnance. To counter this, black stripes were painted across the wing undersides over the six hard-points, to make it harder to see if the bombs had been released. (Terry Hadler)

G1: Brigadier General Thao Ly, 1971

Col. Thao Ly rose to prominence in 1963 when he assumed command of the DNC's airborne regiment. After its disbandment in 1965, Thao Ly was given command of the FAR's second airborne regiment, GM 21. In 1970, he transferred to the guerrilla programme in Savannakhet and was posted in charge of all guerrilla forces in the Third Military Region. During the following year he was promoted to brigadier-general after his participation in retaking the town of Paksong. He then transferred back to the FAR to head its new Second Strike Division. In 1975 he was quickly arrested and executed by the Pathet Lao.

A long-service paratrooper, Gen. Ly wears the FAR red beret with metal airborne badge. Reflecting the persistent French influence in Laos, his jacket and trousers seem to be cut in imitation of the French 1947/56 airborne patterns in 'army green'. His pistol belt, however, is the US M1956. He wears the rank of brigadier-general both on his right collar and on shoulder slides; the FAR General HQ insignia on his left collar; a cloth basic parachutist's wings above his right breast; and the Second Strike Division patch on his left shoulder. He carries an ornate 'swagger stick' as a symbol of authority.

G2: Captain, US Army Special Forces, Project 404, 1972

In 1971 the FAR began to be retrained and reorganised into a light infantry force. Two commando training centres, one at Phou Khou Kouai and a second at Seno, were created for the FAR's new divisions. A third training centre at Champassac was used for two FAR brigades being organised in the Fourth Military Region. US Army assistant military attachés, drawn heavily from the US Army Special Forces, were assigned to Project 404 to act as instructors at the training sites. This captain wears a Special Forces badge and Airborne tab on his left shoulder. Basic US parachutist wings and a CIB are above his left breast. Because Project 404 instructors were usually assigned ostensibly as advisors to service branches of the FAR, a metal Signal Corps badge is on his left shirt pocket. On his right pocket is the cloth insignia for the Fourth Military Region of Laos, worn by those serving at the Champassac training centre. The cloth Laotian wings on his right breast were earned at the FAR airborne training centre at Seno. The 'Laos' tab on his right shoulder was worn in 1971 by those on temporary assignment to Project 404. A green beret with the flash on the 46th Special Forces Company, the parent unit of many Project 404 instructors, is seen hanging from his pants pocket. He carries a 9 mm 'Swedish K' sub-machine gun.

G3: Major, 714ᵉ Bataillon de Parachutistes, 1974

When the Second Strike Division was being

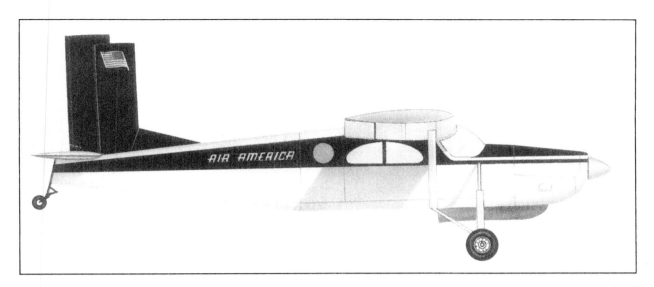

created in 1971, it included a special commando company directly under the command of Gen. Thao Ly. In 1974 the understrength division was renamed 7 Airborne Brigade; the special commando company, known as SPECOM, was expanded to become the 714e Bataillon de Parachutistes of 7 Airborne Brigade. Most of its soldiers were former members of the Savannakhet guerrilla forces; about one-third came from the Commando Raiders. This major, himself a former member of the irregular forces, continues to wear his former guerrilla rank insignia on his right collar. Subdued Laotian wings are worn on his right chest; US wings and a qualification badge from the Thai Rangers are above his left chest, evidence of advanced training in Thailand during his service in the Savannakhet guerrilla programme. Above these is a FAR Combat Infantry Badge, developed in small numbers in 1969 and distributed to élite units. The branch insignia on his left collar, a temple monkey holding a lotus leaf rosette, is for the airborne forces. The silver metal airborne beret badge is a copy of the French equivalent, with the dagger replaced by a Laotian trident; note also the four stars of his present rank. The shoulder insignia is that of 7 Airborne Brigade. He wears US jungle boots, a prized item provided only to SPECOM before 1973. His Vietnamese-style shirt and trousers are made from French-style camouflage material.

Air American Pilatus Porter, 1968. The rugged terrain of north-east Laos made ground transportation impossible in many instances. The government guerrillas relied heavily on transport and resupply aircraft; and many flights were flown by the venerable Pilatus Porter operated by civilian contract airlines. A high-wing monoplane with a steerable tailwheel, it required only 256 feet of runway to land, and 420 feet for take-off. With a lift capacity of more than 1,000 lb and accommodation for five passengers, the Porter was ideally suited for operations in mountainous terrain. The finish was in white and black, with a full-colour US flag on the tail. (Terry Hadler)

H: Laotian and Thai insignia:

H1: Commando Raider Company, Military Region 2, 1969–73.

H2: Groupement Mobile 22, Military Region 2, 1969–73.

H3: Special Guerrilla Programme, Military Region 3, 1967–68.

H4: Guerrilla Forces, Military Region 3, 1968–70.

H5: 1st Special Guerrillà Battalion, Military Region 4, 1968–70.

H6: Commando Raider Company, Military Region 3, 1970–71.

H7: Thai volunteer 'Unity' programme shoulder patch, 1970–73.

H8: Bataillon Commando 615 shoulder patch.

(Acknowledgements to Harry Pugh for assistance with the preparation of this plate.)

INDEX

Figures in **bold** refer to illustrations.